Einstein's Theory of Relativity

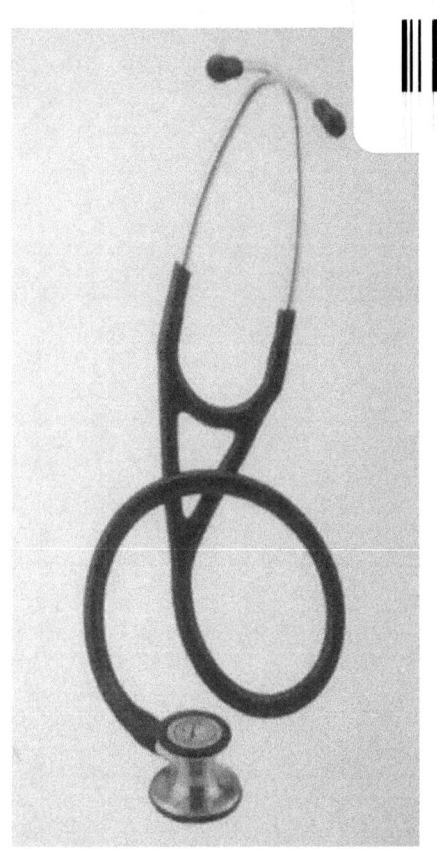

H A Lorentz

EINSTEIN'S THEORY OF RELATIVITY

2019

"Anyone who has never made a mistake has never tried anything new"

6

CONTENTS

Forward by Ruth Finnegan

This concise account of Einstein's theory of relativity was written by his close colleague H A Lorenz and published in 1920 - nearly a century ago, that is, and well before the revolutionary discoveries of the twentieth century to which the theory eventually led.

But despite its date it is still a remarkablepy cogent and readable account of a revolutionary theory, that can assist both initial understanding and, following this, further study and research in this wonderful area of cosmology.

Good wishes in your developing understanding of these new insights into the natural world.

Preface

Whether it is true or not that not more than twelve persons in all the world are able to understand Einstein's Theory, it is nevertheless a fact that there is a constant demand for information about this much-debated topic of relativity. The books published on the subject are so technical that only a person trained in pure physics and higher mathematics is able to fully understand them. In order to make a popular explanation of this far-reaching theory available, the present book is published.

Professor Lorentz is credited by Einstein with sharing the development of his theory. He is doubtless better able than any other man—

except for Einstein   author himself—to explain
this scientific discovery.

Introduction

The action of the Royal Society at its meeting in London on November 6, in recognizing Dr. Albert Einstein's "theory of relativity" has caused a great stir in scientific circles on both sides of the Atlantic. Dr. Einstein propounded his theory nearly fifteen years ago. The present revival of interest in it is due to the remarkable confirmation which it received in the report of the observations made during the sun's eclipse of last May to determine whether rays of light passing close to the sun are deflected from their course.

The actual deflection of the rays that was discovered by the astronomers was precisely what had been predicted theoretically by Einstein many years since. This striking confirmation has led certain German scientists to assert that no scientific discovery of such importance has been made since Newton's theory of gravitation was promulgated. This suggestion, however, was put aside by Dr. Einstein himself when he was interviewed by a correspondent of the New York *Times* at his home in Berlin. To this correspondent he expressed the difference between his conception and the law of gravitation in the following terms:

"Please imagine the earth removed, and in its place suspended a box as big as a room or a whole house, and inside a man naturally floating in the center, there being no force whatever pulling him. Imagine, further, this box being, by a rope or other contrivance, suddenly jerked to one side, which is scientifically termed 'difform motion', as opposed to 'uniform motion.' The person would then naturally reach bottom on the opposite side. The result would consequently be the same as if he obeyed Newton's law of gravitation, while, in fact, there is no gravitation

" ... to every kind of different motion ... "

exerted whatever, which proves that difform motion will in every case produce the same effects as gravitation.

"I have applied this new idea to every kind of difform motion and have thus developed mathematical formulas which I am convinced give more precise results than those based on Newton's theory. Newton's formulas, however, are such close approximations that it was difficult to find by observation any obvious disagreement with experience."

Dr. Einstein, it must be remembered, is a physicist and not an astronomer. He developed his theory as a mathematical formula. The confirmation of it came from the astronomers. As he himself says, the crucial test was supplied by the last total solar eclipse. Observations then proved that the rays of fixed stars, having to pass close to the sun to reach the earth, were deflected the exact amount demanded by Einstein's formulas. The deflection was also in the direction predicted by him.

What is relativity?

The question must have occurred to many, what has all this to do with relativity? When this query was propounded by the *Times* correspondent to Dr. Einstein he replied as follows:

"The term relativity refers to time and space. According to Galileo and Newton, time and space were absolute entities, and the moving systems of the universe were dependent on this absolute time and space. On this conception was built the science of mechanics. The resulting formulas sufficed for all motions of a slow nature; it was found, however, that they would not conform to the rapid motions apparent in electrodynamics.

"This led the Dutch professor, Lorentz, and myself to develop the theory of special relativity. Briefly, it discards absolute time and space and makes them in every instance relative to moving systems. By this theory all phenomena in electrodynamics, as well as mechanics, hitherto irreducible by the old formulae—and there are multitudes—were satisfactorily explained.

"Till now it was believed that time and space existed by themselves, even if there was nothing else—no sun, no earth, no stars—while now we know that time and space are not the vessel for the universe, but could not exist at all if there were no contents, namely, no sun, earth and other celestial bodies.

"This special relativity, forming the first part of my theory, relates to all systems moving with uniform motion; that is, moving in a straight line with equal velocity.

"Gradually I was led to the idea, seeming a very paradox in science, that it might apply equally to all moving systems, even of difform motion, and thus I developed the conception of general relativity which forms the second part of my theory."

As summarized by an American astronomer, Professor Henry Norris Russell, of Princeton, in the *Scientific American* for November 29, Einstein's contribution amounts to this:

"The central fact which has been proved—and which is of great interest and importance—is that the natural phenomena involving gravitation and inertia (such as the motionof the planets) and the phenomena involving electricity and magnetism (including the motion of light) are not independent of one another, but are intimately related, so that both sets of phenomena should be regarded as parts of one vast system, embracing all Nature. The relation of the two is, however, of such a character that it is perceptible only in a very few instances, and then only to refined observations."

Already before the war, Einstein had immense fame among physicists, and among all who are interested in the philosophy of science, because of his principle of relativity.

Clerk Maxwell had shown that light is electro-magnetic, and had reduced the whole theory of electro-magnetism to a small number of equations, which are fundamental in all subsequent work. But these equations were entangled with the hypothesis of the ether, and with the notion of motion relative to the ether. Since the ether was supposed to be at rest, such motion was indistinguishable from absolute motion. The motion of the earth relatively to the ether should have been different at different points of its orbit, and measurable phenomena should have resulted from this difference. But none did, and all attempts to detect effects of motions relative to the ether failed. The theory of relativity succeeded in accounting for this fact. But it was necessary incidentally to throw over the one universal time, and substitute local times attached to moving bodies and varying according to their motion. The equations on which the theory of relativity is based are due to Lorentz, but Einstein connected them with his general principle, namely, that there must be nothing, in observable phenomena, which could be attributed to absolute motion of the observer.

In orthodox Newtonian dynamics the principle of relativity had a simpler form, which did not require the substitution of local time for general time. But it now appeared that Newtonian dynamics is only valid when we confine ourselves to velocities much less than that of light.

The whole Galileo-Newton system thus sank to the level of a first approximation, becoming progressively less exact as the velocities concerned approached that of light.

Einstein's extension of his principle so as to account for gravitation was made during the war, and for a considerable period our astronomers were unable to become acquainted with it, owing to the difficulty of obtaining German printed matter. However, copies of his work ultimately reached the outside world and enabled people to learn more about it. Gravitation, ever since Newton, had remained isolated from other forces in nature; various attempts had been made to account for it, but without success. The immense unification effected by electro-magnetism apparently left gravitation out of its scope. It seemed that nature had presented a challenge to the physicists which none of them were able to meet.

At this point Einstein intervened with a hypothesis which, apart altogether from subsequent verification, deserves to rank as one of the great monuments of human genius. After correcting Newton, it remained to correct Euclid, and it was in terms of non-Euclidean geometry that he stated his new theory. Non-Euclidean geometry is a study of which the primary motive was logical and philosophical; few of its promoters ever dreamed that it would come to be applied in physics. Some of Euclid's axioms were felt to be not "necessary truths," but mere empirical laws; in order to establish this view, self-consistent geometries were constructed upon assumptions other than those of Euclid. In these geometries the sum of the angles of a triangle is not two right angles, and the departure from two right angles increases as the size of the triangle increases. It is often said that in non-Euclidean geometry space has a curvature, but this way of stating the matter is misleading, since it seems to imply a fourth dimension, which is not implied by these systems.

Einstein supposes that space is Euclidean where it is sufficiently remote from matter, but that the presence of matter causes it to become slightly non-Euclidean—the more matter there is in the neighborhood, the more space will depart from Euclid. By the help of this hypothesis, together with his previous theory of relativity, he deduces gravitation—very approximately, but not exactly, according to the Newtonian law of the inverse square.

The minute differences between the effects deduced from his theory and those deduced from Newton are measurable in certain cases. There are, so far, three crucial tests of the relative accuracy of the new theory and the old.

(1) The perihelion of Mercury shows a discrepancy which has long puzzled astronomers. This discrepancy is fully accounted for by Einstein. At the time when he published his theory, this was its only experimental verificationl

(2) Modern physicists were willing to suppose that light might be subject to gravitation—i.e., that a ray of light passing near a great mass like

the sun might be deflected to the extent to which a particle moving with the same velocity would be deflected according to the orthodox theory of gravitation. But Einstein's theory required that the light should be deflected just twice as much as this. The matter could only be tested during an eclipse among a number of bright stars. Fortunately a peculiarly favourable eclipse occurred last year. The results of the observations have now been published, and are found to verify Einstein's prediction. The verification is not, of course, quite exact; with such delicate observations that was not to be expected. In some cases the departure is considerable. But taking the average of the best series of observations, the deflection at the sun's limb is found to be 1.98″, with a probable error of about 6 per cent., whereas the deflection calculated by Einstein's theory should be 1.75″.

It will be noticed that Einstein's theory gave a deflection twice as large as that predicted by the orthodox theory, and that the observed deflection is slightly *larger* than Einstein predicted. The discrepancy is well within what might be expected in view of the minuteness of the measurements. It is therefore generally acknowledged by astronomers that the outcome is a triumph for Einstein.

(3) In the excitement of this sensational verification, there has been a tendency to overlook the third experimental test to which Einstein's theory was to be subjected. If his theory is correct as it stands, there ought, in a

gravitational field, to be a displacement of the lines of the spectrum towards the red. No such effect has been discovered. Spectroscopists maintain that, so far as can be seen at present, there is no way of accounting for this failure if Einstein's theory in its present form is assumed. They admit that some compensating cause *may* be discovered to explain the discrepancy, but they think it far more probable that Einstein's theory requires some essential modification. Meanwhile, a certain suspense of judgment is called for. The new law has been so amazingly successful in two of the three tests that there must be some thing valid about it, even if it is not exactly right as yet.

Einstein's theory has the very highest degree of aesthetic merit: every lover of the beautiful must wish it to be true. It gives a vast unified survey of the operations of nature, with a technical simplicity in the critical assumptions which makes the wealth of deductions astonishing. It is a case of an advance arrived at by pure theory: the whole effect of Einstein's work is to make physics more philosophical (in a good sense), and to restore some of that intellectual unity which belonged to the great scientific systems of the seventeenth and eighteenth centuries, but which was lost through increasing specialization and the overwhelming mass of detailed knowledge. In some ways our age is not a good one to live in, but for those who are interested in physics there are great compensations.

The total eclipse of the sun of May 29,

resulted in a striking confirmation of the new theory of the universal attractive power of gravitation developed by Albert Einstein, and thus reinforced the conviction that the defining of this theory is one of the most important steps ever taken in the domain of natural science. In response to a request by the editor, I will attempt to contribute something to its general appreciation in the following lines.

For centuries Newton's doctrine of the attraction of gravitation has been the most prominent example of a theory of natural science. Through the simplicity of its basic idea, an attraction between two bodies proportionate to their mass and also proportionate to the square of the distance; through the completeness with which it explained so many of the peculiarities in the movement of the bodies making up the solar system; and, finally, through its universal validity, even in the case of the far-distant planetary systems, it compelled the admiration of all.

But, while the skill of the mathematicians was devoted to making more exact calculations of the consequences to which it led, no real progress was made in the science of gravitation.

It is true that the inquiry was transferred to the field of physics, following Cavendish's success in demonstrating the common attraction between bodies with which laboratory work can be done, but it always was evident that natural philosophy had no grip on the universal power of attraction. While in electric effects an influence exercised by the matter placed between bodies was speedily observed—the starting-point of a new and fertile doctrine of electricity—in the case of gravitation not a trace of an influence exercised by intermediate matter could ever be discovered. It was, and remained, inaccessible and unchangeable, without any connection, apparently, with other phenomena of natural philosophy.

Einstein has put an end to this isolation; it is now well established that gravitation affects not only matter, but also light. Thus strengthened in the faith that his theory already has inspired, we may assume with him that there is not a single physical or chemical phenomenon—which does not feel, although very probably in an unnoticeable degree, the influence of gravitation, and that, on the other side, the attraction exercised by a body is limited in the first place by the quantity of matter it contains and also, to some degree, by motion and by the physical and chemical condition in which it moves.

It is comprehensible that a person could not have arrived at such a far-reaching change of view by continuing to follow the old beaten paths, but only by introducing some sort of new idea. Indeed, Einstein arrived at his theory through a train of thought of great originality. Let me try to restate it in concise terms.

The Earth as a Moving Car

Everyone knows that a person may be sitting in any kind of a vehicle without noticing its progress, so long as the movement does not vary in direction or speed; in a car of a fast express train objects fall in just the same way as in a coach that is standing still. Only when we look at objects outside the train, or when the air can enter the car, do we notice indications of the motion. We may compare the earth with such a moving vehicle, which in its course around the sun has a remarkable speed, of which the direction and velocity during a considerable period of time may be regarded as constant. In place of the air now comes, so it was reasoned formerly, the ether which fills the spaces of the universe and is the carrier of light and of electro-magnetic phenomena; there were good reasons to assume that the earth was entirely permeable for the ether and could travel through it without setting it in motion. So here was a case comparable with that of a railroad coach open on all sides. There certainly should have been a powerful "ether wind" blowing through the earth and all our instruments, and it was to have been expected that some signs of it would be noticed in connection with some experiment or other.

Every attempt along that line, however, has remained fruitless; all the phenomena examined were evidently independent of the motion of the earth. That this is the way they do function was brought to the front by Einstein in his first or "special" theory of relativity. For him the ether does not function and in the sketch that he draws of natural phenomena there is no mention of that intermediate matter.

If the spaces of the universe are filled with an ether, let us suppose with a substance, in which, aside from eventual vibrations and other slight movements, there is never any crowding or flowing of one part alongside of another, then we can imagine fixed points existing in it; for example, points in a straight line, located one meter apart, points in a level plain, like the angles or squares on a chess board extending out into infinity, and finally, points in space as they are obtained by repeatedly shifting that level spot a distance of a meter in the direction perpendicular to it.

If, consequently, one of the points is chosen as an "original point" we can, proceeding from that point, reach any other point through three steps in the common perpendicular directions in which the points are arranged. The figures showing how many meters are comprized in each of the steps may serve to indicate the place reached and to distinguish it from any other; these are, as is said, the "co-ordinates" of these places, comparable, for example, with the numbers on a map giving the longitude and latitude.

Let us imagine that each point has noted upon it the three numbers that give its position, then we have something comparable with a measure with numbered subdivisions; only we now have to do, one might say, with a good many imaginary measures in three common perpendicular directions. In this "system of co-ordinates" the numbers that fix the position of one or the other of the bodies may now be read off at any moment.

This is the means which the astronomers and their mathematical assistants have always used in dealing with the movement of the heavenly bodies. At a determined moment the position of each body is fixed by its three co-ordinates. If these are given, then one knows also the common distances, as well as the angles formed by the connecting lines, and the movement of a planet is to be known as soon as one knows how its co-ordinates are changing from one moment to the other. Thus the picture that one forms of the phenomena stands there as if it were sketched on the canvas of the motionless ether.

Einstein's Departure

Since Einstein has cut loose from the ether, he lacks this canvas, and therewith, at the first glance, also loses the possibility of fixing the positions of the heavenly bodies and mathematically describing their movement—i.e., by giving comparisons that define the positions at every moment. How Einstein has overcome this difficulty may be somewhat elucidated through a simple illustration.

On the surface of the earth the attraction of gravitation causes all bodies to fall along vertical lines, and, indeed, when one omits the resistance of the air, with an equally accelerated movement; the velocity increases in equal degrees in equal consecutive divisions of time at a rate that in this country gives the velocity attained at the end of a second as 981 centimeters (32.2 feet) per second. The number 981 defines the "acceleration in the field of gravitation," and this field is fully characterized by that single number; with its help we can also calculate the movement of an object hurled out in an arbitrary direction. In order to measure the acceleration we let the body drop alongside of a vertical measure set solidly on the ground; on this scale we read at every moment the figure that indicates the height, the only co-ordinate that is of importance in this rectilinear movement.

Now we askwhat would we be able to see if the measure were not bound solidly to the earth, if it, let us suppose, moved down or up with the place where it is located and where we are ourselves. If in this case the speed were constant, then, and this is in accord with the special theory of relativity, there would be no motion observed at all; we should again find an acceleration of 981 for a falling body. It would be different if the measure moved with changeable velocity.

If it went down with a constant acceleration of 981 itself, then an object could remain permanently at the same point on the measure, or could move up or down itself alongside of it, with constant speed. The relative movement of the body with regard to the measure should be without acceleration, and if we had to judge only by what we observed in the spot where we were and which was falling itself, then we should get the impression that there was no gravitation at all. If the measure goes down with an acceleration equal to a half or a third of what it just was, then the relative motion of the body will, of course, be accelerated, but we should find the increase in velocity per second one-half or two-thirds of 981. If, finally, we let the measure rise with a uniformly accelerated movement, then we shall find a greater acceleration than 981 for the body itself.

Thus we see that we, also when the measure is not attached to the earth, disregarding its displacement, may describe the motion of the body in respect to the measure always in the same way—*i.e.*, as one uniformly accelerated, as we ascribe now and again a fixed value to the acceleration of the sphere of gravitation, in a particular case the value of zero.

Of course, in the case here under consideration the use of a measure fixed immovably upon the earth should merit all recommendation. But in the spaces of the solar system we have, now that we have abandoned the ether, no such support.

We can no longer establish a system of co-ordinates, like the one just mentioned, in a universal intermediate matter, and if we were to arrive in one way or another at a definite system of lines crossing each other in three directions, then we should be able to use just as well another similar system that in respect to the first moves this or that way. We should also be able to remodel the system of co-ordinates in all kinds of ways, for example by extension or compression. That in all these cases for fixed bodies that do not participate in the movement or the remodelling of the system other co-ordinates will be read off again and again is clear.

New System or Co-Ordinates

What way Einstein had to follow is now apparent. He must—this hardly needs to be said—in calculating definite, particular cases make use of a chosen system of co-ordinates, but as he had no means of limiting his choice beforehand and in general, he had to reserve full liberty of action in this respect. Therefore he made it his aim so to arrange the theory that, no matter how the choice was made, the phenomena of gravitation, so far as its effects and its stimulation by the attracting bodies are concerned, may always be described in the same way —*i.e.*, through comparisons of the same general form, as we again and again give certain values to the numbers that mark the sphere of gravitation. (For the sake of simplification I here disregard the fact that Einstein desires that also the way in which time is measured and represented by figures shall have no influence upon the central value of the comparisons.)

Whether this aim could be attained was a question of mathematical inquiry. It really was attained, remarkably enough, and, we may say, to the surprise of Einstein himself, although at the cost of considerable simplicity in the mathematical form; it appeared necessary for the fixation of the field of gravitation in one or the other point in space to introduce no fewer than ten quantities in the place of the one that occurred in the example mentioned above.

In this connection it is of importance to note that when we exclude certain possibilities that would give rise to still greater intricacy, the form of comparison used by Einstein to present the theory is the only possible one; the principle of the freedom of choice in co-ordinates was the only one by which he needed to allow himself to be guided.

Although thus there was no special effort made to reach a connection with the theory of Newton, it was evident, fortunately, at the end of the experiment that the connection existed. If we avail ourselves of the simplifying circumstance that the velocities of the heavenly bodies are slight in comparison with that of light, then we can deduce the theory of Newton from the new theory, the "universal" relativity theory, as it is called by Einstein.

Thus all the conclusions based upon the Newtonian theory hold good, as must naturally be required.

But now we have got further along. The Newtonian theory can no longer be regarded as absolutely correct in all cases; there are slight deviations from it, which, although as a rule unnoticeable, once in a while fall within the range of observation.

Now, there was a difficulty in the movement of the planet Mercury which could not be solved. Even after all the disturbances caused by the attraction of other planets had been taken into account, there remained an inexplicable phenomenon—*i.e.*, an extremely slow turning of the ellipsis described by Mercury on its own plane; Leverrier had found that it amounted to forty-three seconds a century. Einstein found that, according to his formulas, this movement must really amount to just that much. Thus with a single blow he solved one of the greatest puzzles of astronomy.

Still more remarkable, because it has a bearing upon a phenomenon which formerly could not be imagined, is the confirmation of Einstein's prediction regarding the influence of gravitation upon the course of the rays of light. That such an influence must exist is taught by a simple examination; we have only to turn back for a moment to the following comparison in which we were just imagining ourselves to make our observations. It was noted that when the compartment is falling with the acceleration of

981 the phenomena therein will occur just as if there were no attraction of gravitation. We can then see an object, *A*, stand still somewhere in open space. A projectile, *B*, can travel with constant speed along a horizontal line, without varying from it in the slightest.

A ray of light can do the same; everybody will admit that in each case, if there is no gravitation, light will certainly extend itself in a rectilinear way. If we limit the light to a flicker of the slightest duration, so that only a little bit, *C*, of a ray of light arises, or if we fix our attention upon a single vibration of light, *C*, while we on the other hand give to the projectile, *B*, a speed equal to that of light, then we can conclude that *B* and *C* in their continued motion can always remain next to each other. Now if we watch all this, not from the movable compartment, but from a place on the earth, then we shall note the usual falling movement of object *A*, which shows us that we have to deal with a sphere of gravitation. The projectile *B* will, in a bent path, vary more and more from a horizontal straight line, and the light will do the same, because if we observe the movements from another standpoint this can have no effect upon the remaining next to each other of *B* and *C*.

Deflection of Light

The bending of a ray of light thus described is much too light on the surface of the earth to be observed. But the attraction of gravitation exercised by the sun on its surface is, because of its great mass, more than twenty-seven times stronger, and a ray of light that goes close by the superficies of the sun must surely be noticeably bent. The rays of a star that are seen at a short distance from the edge of the sun will, going along the sun, deviate so much from the original direction that they strike the eye of an observer as if they came in a straight line from a point somewhat further removed than the real position of the star from the sun.

It is at that point that we think we see the star; so here is a seeming displacement from the sun, which increases in the measure in which the star is observed closer to the sun. The Einstein theory teaches that the displacement is in inverse proportion to the apparent distance of the star from the centre of the sun, and that for a star just on its edge it will amount to 1'.75 (1.75 seconds). This is approximately the thousandth part of the apparent diameter of the sun.

Naturally, the phenomenon can only be observed when there is a total eclipse of the sun; then one can take photographs of neighboring stars and through comparing the plate with a picture of the same part of the heavens taken at a time when the sun was far removed from that point the sought-for movement to one side may become apparent.

Thus to put the Einstein theory to the test was the principal aim of the English expeditions sent out to observe the eclipse of May 29, one to Prince's Island, off the coast of Guinea, and the other to Sobral, Brazil. The first-named expedition's observers were Eddington and Cottingham, those of the second, Crommelin and Davidson. The conditions were especially favorable, for a very large number of bright stars were shown on the photographic plate; the observers at Sobral being particularly lucky in having good weather.

The total eclipse lasted five minutes, during four of which it was perfectly clear, so that good photographs could be taken. In the report issued regarding the results the following figures, which are the average of the measurements made from the seven plates, are given for the displacements of seven stars:

$1''.02, 0''.92, 0''.84, 0''.58, 0''.54, 0''.36, 0''.24$, whereas, according to the theory, the displacements should have amounted to: $0''.88$, $0''.80, 0''.75, 0''.40, 0''.52, 0''.33, 0''.20$.

If we consider that, according to the theory the displacements must be in inverse ratio to the distance from the centre of the sun, then we may deduce from each observed displacement how great the sideways movement for a star at the edge of the sun should have been. As the most probable result, therefore, the number $1''.98$ was found from all the observations together. As the last of the displacements given above—*i.e.,* 0". 24 is about one-eighth of this, we may say that the influence of the attraction of the sun upon light made itself felt upon the ray at a distance eight times removed from its centre.

The displacements calculated according to the theory are, just because of the way in which they are calculated, in inverse proportion to the distance to the centre. Now that the observed deviations also accord with the same rule, it follows that they are surely proportionate with the calculated displacements. The proportion of the first and the last observed sidewise movements is 4.2, and that of the two most extreme of the calculated numbers is 4.4.

This result is of importance, because thereby the theory is excluded, or at least made extremely improbable, that the phenomenon of refraction is to be ascribed to, a ring of vapor surrounding the sun for a great distance. Indeed, such a refraction should cause a deviation in the observed direction, and, in order to produce the displacement of one of the stars under observation itself a slight proximity of the vapor ring should be sufficient, but we have every reason to expect that if it were merely a question of a mass of gas around the sun the diminishing effect accompanying a removal from the sun should manifest itself much faster than is really the case.

We cannot speak with perfect certainty here, as all the factors that might be of influence upon the distribution of density in a sun atmosphere are not well enough known, but we can surely demonstrate that in case one of the gasses with which we are acquainted were held in equilibrium solely by the influence of attraction of the sun the phenomenon should become much less as soon as we got somewhat further from the edge of the sun. If the displacement of the first star, which amounts to 1.02-seconds were to be ascribed to such a mass of gas, then the displacement of the second must already be entirely inappreciable.

So far as the absolute extent of the displacements is concerned, it was found somewhat too great, as has been shown by the figures given above; it also appears from the final result to be 1.98 for the edge of the sun—*i.e.,* 13 per cent, greater than the theoretical value of

1.75. It indeed seems that the discrepancies may be ascribed to faults in observations, which supposition is supported by the fact that the observations at Prince's Island, which, it is true, did not turn out quite as well as those mentioned above, gave the result, of 1.64, somewhat lower than Einstein's figure.

(The observations made with a second instrument at Sobral gave a result of 0.93, but the observers are of the opinion that because of the shifting of the mirror which reflected the rays no value is to be attached to it.)

Difficulty Exaggerated

During a discussion of the results obtained at a joint meeting of the Royal Society and the Royal Astronomical Society held especially for that purpose recently in London, it was the general opinion that Einstein's prediction might be regarded as justified, and warm tributes to his genius were made on all sides. Nevertheless, I cannot refrain, while I am mentioning it, from expressing my surprise that, according to the report in *The Times* there should be so much complaint about the difficulty of understanding the new theory. It is evident that Einstein's little book

"About the Special and the General Theory of Relativity in Plain Terms," did not find its way into England during wartime. Any one reading it will, in my opinion, come to the conclusion that the basic ideas of the theory are really clear and simple; it is only to be regretted that it was impossible to avoid clothing them in pretty involved mathematical terms, but we must not worry about that.

I allow myself to add that, as we follow Einstein, we may retain much of what has been formerly gained. The Newtonian theory remains in its full value as the first great step, without which one cannot imagine the development of astronomy and without which the second step, that has now been made, would hardly have been possible. It remains, moreover, as the first, and in most cases, sufficient, approximation.

It is true that, according to Einstein's theory, because it leaves us entirely free as to the way in which we wish to represent the phenomena, we can imagine an idea of the solar system in which the planets follow paths of peculiar form and the rays of light shine along sharply bent lines—think of a twisted and distorted planetarium—but in every case where we apply it to concrete questions we shall so arrange it that the planets describe almost exact ellipses and the rays of light almost straight lines.

It is not necessary to give up entirely even the ether. Many natural philosophers find satisfaction in the idea of a material intermediate substance in which the vibrations of light take place, and they will very probably be all the more inclined to imagine such a medium when they learn that, according to the Einstein theory, gravitation itself does not spread instantaneously, but with a velocity that at the first estimate may be compared with that of light. Especially in former years were such interpretations current and repeated attempts were made by speculations about the nature of the ether and about the mutations and movements that might take place in it to arrive at a clear presentation of electro-magnetic phenomena, and also of the functioning of gravitation.

In my opinion it is not impossible that in the future this road, indeed abandoned at present, will once more be followed with good results, if only because it can lead to the thinking out of new experimental tests. Einstein's theory need not keep us from so doing; only the ideas about the ether must accord with it.

Nevertheless, even without the color and clearness that the ether theories and the other models may be able to give, and even, we can feel it this way, just because of the soberness induced by their absence, Einstein's work, we may now positively expect, will remain a monument of science; his theory entirely fulfills the first and principal demand that we may make, that of deducing the course of phenomena from certain principles exactly and to the smallest details. It was certainly fortunate that he himself put the ether in the background; if he had not done so, he probably would never have come upon the idea that has been the foundation of all his examinations.

Thanks to his indefatigable exertions and perseverance, for he had great difficulties to overcome in his attempts, Einstein has attained the results, which I have tried to sketch, while still young; he is now 45 years old. He completed his first investigations in Switzerland, where he first was engaged in the Patent Bureau at Berne and later as a professor at the Polytechnic in Zurich. After having been a professor for a short time at the University of Prague, he settled in Berlin, where the Kaiser Wilhelm Institute afforded him the opportunity to devote himself exclusively to his scientific work. He repeatedly visited our country and made his Netherland colleagues, among whom he counts many good friends, partners in his studies and his results. He attended the last meeting of the department of natural philosophy of the Royal Academy of Sciences, and the members then had the privilege of hearing him explain, in his own fascinating, clear and simple way, his interpretations of the fundamental questions to which his theory gives rise.

THE END

but the theory continues to inspire and to be built on (perhaps by you)

Appendix

Some quotes attributed to Einstein p that might inspire you:

Insanity: doing the same thing over and over again and expecting different results.

Imagination is more important than knowledge.

If you can't explain it simply, you don't understand it well enough.

Two things are infinite: the universe and human stupidity; and I'm not sure about the universe.

Life is like riding a bicycle. To keep your balance you must keep moving.

No problem can be solved from the same level of consciousness that created it.

The important thing is not to stop questioning. Curiosity has its own reason for existing.

I have no special talents. I am only passionately curious.

Anyone who has never made a mistake has never tried anything new.

Logic will get you from A to B. Imagination will take you everywhere.

More things you might like

Walter Isaacson, *Einstein: His Life and Universe,* Simon and Schuster, 2017"

Martin Rees, *From Here to Infinity: Scientific Horizons*, Profile Books, 2011.

Martin Rees, *Just* Six Numbers! W&N, 2015.

Questions and projects for discussion, debate or action

These are designed to help you towards

- understanding a key notion of modern times
- further investigating its content, impact and (in the light of more recent discoveries) its limitations
- setting it in the context of earlier and current work in cosmology
- Increasing your insight into how science works
- making this understanding personal to yourself and perhaps your life plans

What further discoveries or insights did Einstein's theory of relativity give rise to? (you can check this out easily on the web)

In your opinion which if any of the following do these new theories belong to?

philosophy

physics

mythology and symbolism

astrophysics

cosmology

astronomy

astrology

poetry

mathematics

empirical science

several of the above

all of the above

Are they revolutionary in the same sense as Einstein's theory? Why, or why not?

The theory of relativity has often been compared to Newton's theory of gravity. In your opinion does it contradict, complement or build on that theory?

Could it have happened without it and how does it differ?

Do you think there is just one universe or several? Why do you think this?

Building on this account and on your knowledge or, or researches into, subsequent discoveries / theories , would you say that science progresses by revolutionary paradigm changes (as Kuhn has it) or little by little, incrementally?

What are Black Holes and the Big Bang and why are they not mentioned in the account here? How much do you know about them?

What most surprised you in it?

And -

   Would you like to be a physicist or cosmologist yourself? Why/why not?

****

Join the Hearing Others' Voices Facebook community

https://m.facebook.com/HearingOthers/

You are welcome - make your voice heard!

94

About the 'Hearing Others' Voices' series

The series *(not* exam or curriculum related) aims to provide young adults with accessibly written introductions to topics, insights, and voices beyond their usual experience.

Hearing Others' Voices (https://
www.balestier.com/category/hearing-others-
voices/) is a transcultural and transdisciplinary
book series edited by British anthropologist Ruth
Finnegan and Taiwanese physicist Roh-Suan
Tung. The books are designed, in straightforward
language, to enlighten and attract general
readers, undergraduates, young adults and,
above all, teenagers about recent advances in
thought, overlooked areas of the world, and key
issues of the day.

They tackle perpetual topics of interest such as
mental health, the nature of the universe, of
music and of storms, shamanism, voices of the
Christian west, and our awesome minds and
bodies. They provide introductions, accessibly
written (often illustrated, and linked to audio-
video internet material), to the latest insights into
fascinating and perpetual topics such as the
different ways pain has been handled in the past

and present, the history of children, the nature of poetry, and more. They attempt, too, to reveal something of the wisdom of the often dismissed traditions of, for example, hunters and gatherers, Amazonia, native Americans, and Africa, bringing a deeper and more balanced perspective on who we are, and how the world has come to be as it is.

The first volumes to be released (preceded by the October launch of the Chinese version of Rob Janoff's amazing personal account) were:

Taking a bite out off the apple, a graphic designer's tale, Rob Janoff,   the world's most famous logo designer

Time for the world to learn from Africa, Ruth Finnegan, anthropologist,

Voices from the Christian west,   Venerable Archdeacon Arthur Hawes, theologian, Emeritus Archdeacon, Lincoln Cathedra

Listen world! Evelyn Glennie, the workf's premier solo percussionist

Native American knowledge systems: something to learn, Clara Sue Kidsell, Emeritus Professor, University of North Carolina

From blank canvas to garment: a creative journey of discovery, Marella Campagna, fashion designer

. Other early volumes (2019-20) include
For peace. Voices against the fog and blood of war, edited by Ruth Finnegan, anthropologist
Decisions, decisions, decisions, with compassion   and love: voice from the headteacher's study, A headteacher (anonymous)
Being a poet, being an Archbishop, being human,   Rowan Williams, writer, poet, theologian, previously Archbishop of Canterbury

Further books will appear over the coming years, many also with translations into Chinese and other world languages.

www.ingramcontent.com/pod-product-compliance
Lightning Source LLC
Chambersburg PA
CBHW061148180526
45170CB00002B/672